HOW TO USE THIS BOOK

Laminate may be the perfect flooring material for you. It's a beefed-up version of the plastic counter surfaces you find in kitchens, the stuff is both solid and strong. It installs over existing flooring in many cases and is pretty easy to do. However, it can cost as much as a solid wood floor but you can't refinish it if it gets damaged.

Whether it's right for you depends on your lifestyle, your budget, and where you want to install it.

To help you decide, this book tells you all about it. The first section *(pages 2 to 9)* is about the flooring itself. The second section *(pages 10 to 13)* tells you what you need to install it, and the last section *(pages 14 to 31)* tells you how to do it.

Take notes, jot down ideas or questions, and work out your plan. Then take the book to your flooring dealer and put your plan into action.

Laminate planks, glued together, resting on the floor but not nailed to it

Underlayment for cushioning and sound reduction

Vapor barrier protects against moisture from concrete subfloor; sometimes not needed on wood subfloors

Existing flooring or subfloor

Looks a lot like wood, doesn't it?
But it isn't. It's a reasonably new product on the market called laminate flooring. It comes in a variety of colors and patterns that imitate wood, stone, or ceramic tile.

1

THE FLOORING MATERIAL

All laminate flooring is made essentially the same way. The top wear layer is cellulose paper impregnated with clear melamine resins. Just below it is the design layer—a photo or pattern printed on paper and strengthened with resins. The core is usually a durable fiberboard.

The bottom stabilizing layer is made of paper or melamine. *(See page 3 for details.)*

Individual laminate pieces can look like real wood or stone. There will be pattern repeats in the floor, though; something you won't find in a flooring made from natural wood or stone.

FLOORING UNDER PRESSURE

Laminate flooring differs in more than just color and size. There are two distinct construction methods. With high-pressure lamination, the bottom and top layers are each heated and pressurized into laminate structures. These layers are then fused to the core with glue under heat and pressure. With direct-pressure construction, the layers are assembled all at once, then filled with hardening melamine resins using heat and pressure. High-pressure types are more impact- and dent-resistant. Direct-pressure laminates are more economical and offer very good quality overall. The packaging should tell you which kind is inside; if it doesn't, ask the salesperson or call the manufacturer.

PLANKS

Planks are meant to look like a variety of natural materials, including stone and wood. Planks that imitate wood come in a variety of "species," colors, and patterns. Plank sizes may vary between manufacturers, but not by much. They're all about 8 inches wide and about 4 feet long. They have tongue-and-groove edges that you join with glue.

TILES

Tiles usually mimic other kinds of flooring, such as ceramic tile or stone. They can be either individual squares or larger squares with imitation grout joints. Like planks, they're fitted tongue-to-groove and glued.

IF IT'S NOT WOOD, WHAT IS IT?

It's a combination of layers that form a solid, long-lasting flooring material. Manufacturers have different names for these layers. Some combine the design and wear layers, but the basics are the same.

Wear layer
Made of clear melamine resins combined with cellulose paper, this layer resists dents, scratches, burns, and fading.

Core layer
Made of durable fiberboard and saturated with resins for durability.

Design layer
Printed pattern or photo in this layer makes the laminate look like natural materials. Melamine resins add strength.

Stability or balancing layer
Made of polymer laminated paper. Adds stability, allowing planks to adapt to temperature and humidity changes without warping.

COMPARING FLOORING OPTIONS

Still trying to decide if laminate flooring is for you? The chart below compares laminate with wood and vinyl flooring. Add your own concerns in the space provided at the bottom and ask your flooring dealer to fill in the details. Don't be afraid to shop around.

	LAMINATE	WOOD	VINYL
Cost (per roughly 400 square feet)	$1,800-$2,800	$1,800-$2,900	$300-$1,900
Durability	Warp-, rip-, scratch-resistant. Damaged planks replaceable.	Can be refinished. Lasts indefinitely.	May rip, bend, scratch. Easy to replace.
Impact resistance (for denting and cracking)	Varies with quality. Resists pressure of at least 4,250 lbs. per square inch. Some resist up to 9,000 lbs.	Varies a lot between species. Oak is most resistant. White pine dents more easily.	Resists pressure of up to 200 lbs. per square inch.
Color	Batches always match.	Varies from tree to tree; is part of its charm.	Varies from batch to batch.
Stain resistance	High. Made of waterproof resins.	Low. Absorbs water. Finish is its only protection.	Medium.
UV resistance	High. Won't fade.	Low. May fade or darken with age.	Low. May fade.
Ease of installation	Easy. Material adapts to some irregularity in floor.	Easy floor preparation.	Subfloor preparation is difficult. Irregularities and dirt may show in finish flooring.
Wear layer warranty	Up to 15 years is common.	No warranty, but can be refinished.	Up to 10 years limited warranty.
Questions for dealer			

THE FLOORING SYSTEM

Laminate flooring planks are glued to each other; they're not nailed, stapled, glued, or otherwise fastened down. They "float" on the subfloor, held in place by gravity. Even the underlayment and vapor barrier, if there is one, are loose. When the flooring swells and shrinks with changes in humidity, it will move as a unit. The movement is imperceptible, but enough to ensure the floor won't warp or buckle. To accommodate expansion, leave a 1/4-inch gap between the edge of the flooring and the walls; the shoe molding will hide it.

You can install laminate right over many existing surfaces. Flip ahead to page 15 for details.

Always use the specific products, glues, and methods your flooring manufacturer recommends. Substitutes may void your warranty.

COVERING WOOD

Wood subfloors usually need only underlayment— even if they're covered with some other type of flooring. Solid panels (right) are butted and loosely taped. You could also use foam, cork, or any other underlayment that your flooring manufacturer recommends— but only one type per floor. Solid panels, though more expensive than the others, do the best job of soundproofing.

Flip ahead to page 15 for details.

ACCLIMATE THE FLOORING

Even after it's installed, laminate flooring will expand and contract with changes in temperature and humidity. To keep those changes to a minimum in the installed floor, it's important to let the newly bought flooring adjust to its new climate conditions. Lay the flooring flat in its original, unopened cartons and leave it in the room where it'll be installed for at least 48 hours before you install it.

Laminate planks
Joints glued tongue-to-groove.

Joists

Solid underlayment
Panels loosely butted and taped. Deadens sound, gives heat insulation, and improves walking comfort.

Plywood subfloor
Bare or covered with other flooring. Cover with vapor barrier if flooring manufacturer recommends it.

COVERING CONCRETE

Because concrete always gives off a bit of moisture, do a moisture test to see if your floor is suitable for laminate (page 17). Either solid underlayment or foam may be used, but foam is less expensive. Always use a vapor barrier when you're installing flooring over concrete, even if there's a surface floor such as ceramic tile on top. Remove any wood glued to concrete. Check the manufacturer's specifications for other restrictions.

Rolled foam underlayment
Flexible, closed-cell foam sold in rolls. Rolls butt against the walls and against each other.

Laminate planks
Joints glued tongue-to-groove.

Concrete subfloor
Bare or covered with surface flooring.

Vapor barrier
Overlapping plastic sheets protect against normal moisture levels. Required with concrete subfloor.

THE EXPANSION GAP

Whether you install the flooring over a wood subfloor, a concrete slab, or existing flooring, you'll need to leave a gap at the walls for expansion. Recommendations vary, but a 1/4-inch space is typical. Maintain the right gap with temporary spacers sold by the flooring manufacturer.

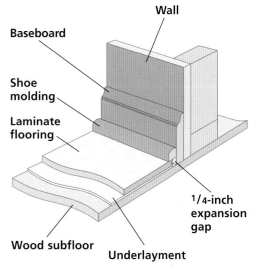

Wall

Baseboard

Shoe molding

Laminate flooring

1/4-inch expansion gap

Wood subfloor

Underlayment

MAYBE YOU SHOULD MEASURE IN METRIC

Laminate flooring has been popular in Europe for years, and many manufacturers sell their flooring in that market as well as in North America. As a result, some planks have odd sizes when measured in inches, but are much easier to handle in metric. Don't panic. You won't need to learn a whole new system. All you really need to do is get a tape measure with both standard and metric measurements printed on it. When it's easier to work in metric, use the metric side. When it's more comfortable to work in inches, work in inches.

All laminate floors are cushioned by an underlayment that deadens the sound, cushions the step, and keeps the glue from fastening the planks to the subfloor during installation. Concrete subfloors always require a vapor barrier under the underlayment to keep normal moisture from the concrete from damaging the laminate floor. Many manufacturers provide a choice of underlayments and vapor barriers. Here's a guide to what's available:

Vapor barrier
Polyethylene film comes in different thicknesses (about 6-8 mil) from different manufacturers. Keeps moisture from concrete from seeping into floor. Cover with underlayment.

Foam
Closed-cell polyethylene that cushions the floor and diminishes noise. Comes in rolls.

2-in-1 foam
Combination vapor barrier/underlayment that lets you cover concrete subfloors with one layer instead of two. Place it film-side down. Check with your dealer for availability and directions.

Solid and rubber-backed polyester
Available in boards or rolls, respectively. Produces the most comfortable floor and deadens sound more than foam does, but is more expensive.

COVERING STAIRS

The rules for stairs are different. In this case only, you glue the flooring down; it doesn't "float." Don't bother using a vapor barrier or underlayment here, either. Cover the treads, risers, and exposed tread edges with planks cut to fit. Glue and screw edging pieces to each tread. You may need to trim the last flooring plank to make room for the first nosing and the 1/4-inch gap between the nosing and the floor. It's easiest to plan for this as you install your floor. Be sure to follow the manufacturer's directions carefully.

WISDOM OF THE AISLES

Laminate flooring can look good on enclosed staircases when either the floor above or the floor below is also covered with laminate. If you're covering the upper floor with laminate, make sure you leave room for the nosing at the top of the stairs. If you're covering the bottom floor with laminate, glue shoe molding at the foot of the bottom riser. Let a pro take care of staircases that are open at one side. The exposed sides of the treads and cut edges of laminate are hard to handle.

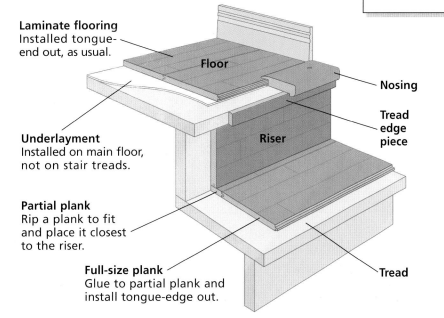

Laminate flooring
Installed tongue-end out, as usual.

Floor

Nosing

Tread edge piece

Underlayment
Installed on main floor, not on stair treads.

Riser

Partial plank
Rip a plank to fit and place it closest to the riser.

Full-size plank
Glue to partial plank and install tongue-edge out.

Tread

ALLERGY SUFFERERS REJOICE!

Because it's made of large, solid, smooth pieces with tightly butted joints, laminate flooring offers allergy sufferers a bonus: no cracks to trap dust and no fibers to hold allergens. And unlike some other types of flooring, laminate planks emit very few gases as they age.

FINISHING UP

Laminate trim pieces add a final decorative touch. Most manufacturers offer a selection of pieces in colors and patterns to match their flooring. A range of common options are shown below. Some are almost essential. If you're installing two different laminate floors, a laminate T-molding designed to fit the gap between them is probably your best bet. Other trim is optional—a traditional shoe molding will cover the expansion gap as well as a laminate one. Whatever trim pieces you choose, let the floor dry overnight before you install them.

MAINTAINING THE GAP

Trim pieces, like reducer strips and T-moldings, cover the transition between two different floors. The molding is designed around a metal track that maintains the gap. You screw the track to the subfloor, leaving the proper gap between the track and the laminate floor. The trim snaps into the track.

DECORATIVE GRATES

Finish off your new wood-look laminate floor with a new real-wood vent *(below)*. They're available in a range of wood species and stains to complement laminate floors.

LAMINATE TRIM OPTIONS

Shoe molding
Nail to baseboard to cover the expansion gap at the edge of the flooring.

Stair nosing
Overhangs steps. Screw and glue in place.

Baseboard
Matches or contrasts with flooring. Nailed to the wall studs.

T-molding
Handles transition between two different laminate floors or floors of same height.

End molding
Finishes edge of flooring where shoe molding is not possible.

Reducer strip
Handles transition from laminate floor to lower floor.

DESIGN DECISIONS

Designing with laminate flooring is easy. Pieces from the same manufacturer will all be the same size and shape, and the tongues and grooves will all fit together. You can install nearly any pattern you can dream up. So all you have to do is decide: Do you prefer the traditional look of wood flooring, or the look of bold, colorful patterns? Do you want to accent one area with patterned panels, or mix colors across the whole floor? It's up to you.

PLACING THE PLANKS

Laminate flooring manufacturers specify the way to lay planks, based on room shape, focal point, and length of starting wall. Generally, lay planks parallel to incoming sources of light *(left)*. Hallways and other narrow areas usually look best with planks oriented along the length of the space *(right)*.

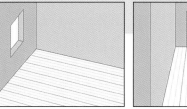

COLOR & PATTERN

Light colors and simple patterns can make a room seem brighter and more open. Darker colors and prominent patterns, on the other hand, tend to make a large space seem smaller and more intimate.

MIX & MATCH

Take a playful approach to your floor. Create a checkerboard (right) with different colors. Accent a hallway or entrance by changing the pattern.

9

CALCULATING FLOORING NEEDS

To know how much flooring to buy, all you really need to know is the square footage of the space to be covered. And that's easy to calculate—it's just length times width, right? Well, maybe not, if your room isn't a simple square or rectangle.

For complicated shapes, draw a plan like the one at right. Divide the room into sections based on simple shapes like squares, triangles, and circles. Then measure each separately, write your results in the box, and add them up. Add 10 percent for installation errors and to have a few planks left over for future repairs.

Write a list of any new trim pieces you want and mark on your plan where they'll go.

CALCULATING SQUARE FOOTAGE

Take all measurements in inches; calculate the area as follows:

1 Rectangles (**A**): Multiply length by width.
2 Right triangles (**B** and **C**): Multiply length by width and divide the answer by 2.
3 Circles (**D**): For full circles, multiply 3.14 (π) by the radius squared (r^2). Divide this by 2 for half circles, by 4 for quarter circles, etc.
4 Total everything and divide by 144 to convert to square feet. Add 10% for waste. Round up to the next whole number.

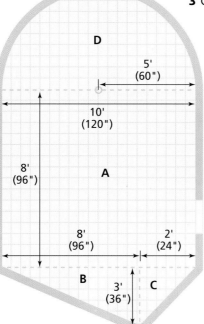

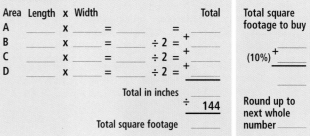

EXAMPLE

Area	Length	x	Width				Total
A	96"	x	120"	=	11,520	=	11,520
B	36"	x	96"	=	3,456	÷ 2 =	+ 1,728
C	36"	x	24"	=	864	÷ 2 =	+ 432
D	π (3.14)	x	60"²	=	11,304	÷ 2 =	+ 5,652

Total in inches 19,332
÷ **144**

Total square footage 134.25

Total square footage to buy

134.25
(10%) + 13.43

147.68

Round up to next whole number 148

YOUR ROOM

Area	Length	x	Width				Total
A	_____	x	_____	=	_____	=	_____
B	_____	x	_____	=	_____	÷ 2 =	+ _____
C	_____	x	_____	=	_____	÷ 2 =	+ _____
D	_____	x	_____	=	_____	÷ 2 =	+ _____

Total in inches _____
÷ **144**

Total square footage _____

Total square footage to buy

_____
(10%) + _____

_____

Round up to next whole number _____

TOOLS FOR LAMINATE FLOORING

Most of the tools you'll need to install laminate flooring are pretty basic. You probably already have most of them. Any you don't have you can find at your local hardware store. You might consider renting some of the more expensive ones that you won't use much.

You'll need some special tools, such as tapping blocks and clamps. They're product-specific; buy or rent them from your flooring dealer.

To replace a badly damaged plank, you'd need some other tools. It's usually a pretty tricky job, though, so hire a professional installer.

SAFETY COMES FIRST

Even though installing laminate flooring is easy, it's still easy to get hurt if you don't take the proper precautions. Wear safety goggles when using a hammer or other striking tool. Wear rubber gloves when handling caustic materials, such as leveling compound. Wear a dust mask when sanding. Good quality knee pads will make the whole job more comfortable.

PREPARATION TOOLS

1. Pry bar
For removing shoe molding and other trim.

2. Carpenter's level
To check subfloor for level and high/low spots.

3. Cold chisel
Works as a wedge to pry up old wood floor.

4. Ball-peen hammer
Heavy, large-faced hammer for hitting cold chisel; could substitute claw hammer.

5. Large putty knife
Use a metal knife for prep work. Use only a plastic scraper on floor surface.

6. Flat-head screwdriver
Wedge it under shoe moldings to loosen them.

7. Trowel
Smooths leveling compound.

8. Jamb saw
Cuts door jamb and door trim to create clearance for flooring.

9. Belt sander with 60-grit paper
For smoothing a wood subfloor.

10. Floor scraper
Removes residue from subfloor.

INSTALLATION TOOLS

1. Backsaw with miter box
For cutting angles.

2. Circular saw
For making straight cuts. Use a carbide-tipped blade.

3. Saber saw
For cutting curves. Use a laminate blade.

4. Electric drill
For drilling holes.

5. Tape measure
For accurate measurements.

6. Plastic putty knife
Apply putty with plastic knife to avoid scratching laminate.

7. Compass
Trace along wall to transfer irregularities to floor.

8. China marker
Mark holes and scribe lines.

9. Utility knife
For cutting underlayment.

10. Phillips-head screwdriver

11. Coping saw
Cutting coping joints in molding.

12. Hacksaw
For cutting metal track to install laminate moldings.

13. Flexi-curve
Form around odd-shaped obstacles, then trace onto planks.

14. Combination square
Measuring and marking cuts on planks.

15. Nail set
Countersinking nail heads in shoe molding.

16. Claw hammer
Fastening shoe moldings.

17. Caulking gun
Applying silicone caulk.

18. Power miter saw
Makes short work of angled cuts on trim.

SPECIALIZED ITEMS

1. Installation clamp
Keeps rows tight during installation. May come with extension straps for clamping large areas.

2. Installation strap
Extend across floor to keep rows straight and tight; use two to three per plank length.

3. Pull bar
Slip in place and tap to tighten joints.

4. Spacers
For maintaining expansion space around walls and other solid objects.

5. Tapping block
Slit in side of block fits over tongue on plank; tap gently to tighten joint.

12

GLUE

Glue is crucial to laminate flooring. Glue between the flooring pieces is the only thing keeping the floor together. Each manufacturer makes its own glue—don't substitute, or you may void your floor's warranty. Different manufacturers require glue to be applied to different parts of the board, as discussed below. Read the instructions that come with your flooring to know how much glue to use and where to put it. Some manufacturers even specify how to cut open the bottles of glue.

WISDOM OF THE AISLES

Cleaning up fresh glue that squeezes up as you lay the flooring is simply a matter of wiping the surface with a clean damp cloth. If you miss a bit, don't worry—even dried glue usually comes off with a damp cloth. If plain water doesn't work, try acetone. Apply a bit to a soft cloth and gently rub off the glue. Don't use floor cleaners other than those supplied by the laminate flooring manufacturer.

WHERE TO PUT IT

Where you're supposed to put glue depends on the product you buy. You might be instructed to put glue only on the flooring tongue *(top)*, in the groove *(bottom)*, or in both places. Follow the instructions.

If floor seams overlap with those of solid underlayment, fully tape the underlayment seams to keep glue from dripping through.

HOW MUCH IS ENOUGH?

The minute the job starts looking neat and tidy, you know you're not putting in enough glue. The glue must squeeze out of the joints when you clamp the boards together. If it doesn't, small gaps will develop between the planks as the glue dries. The amount of glue you use depends on how tightly the tongue fits into its groove. Tight fits require less glue *(top)*; loose fits require more *(bottom)*.

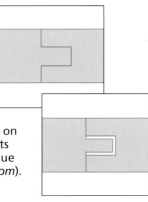

PREPARING THE ROOM

LAST-MINUTE CHECKLIST

Make sure you have:

- ✓ Utility knife
- ✓ Screwdriver
- ✓ Shim
- ✓ Pry bar
- ✓ Jamb saw
- ✓ Plank
- ✓ Underlayment and vapor barrier
- ✓ 1x2
- ✓ Electric drill

Preparing to install laminate flooring is pretty easy because you can usually install it right over the existing flooring as long as it has been cleaned thoroughly. There are some exceptions, of course, so follow the manufacturer's directions. The chart on page 15 is a good general guide.

Begin by removing the shoe moldings. Remove baseboards if you want to replace them with laminate versions. Cut the door jamb and the trim so the flooring will fit underneath. If you'll be installing a finishing strip, such as a reducer, you may need to install a temporary backstop in the doorway. Previously installed wood flooring over concrete usually has to be removed.

Removing shoe moldings
Break the paint seal between shoe molding and baseboard with a utility knife. Wedge a slotted screwdriver under the molding to pop it up, then pry it out with a pry bar. Protect the baseboard with a shim.

Trimming door frames
Lay a floor plank upside down on top of a piece of the underlayment and vapor barrier. Cut off the bottom of the jamb and casing with a jamb saw. The plank should slide easily underneath.

Fastening a 1x2 backstop
Drill pilot holes for three $2\frac{1}{4}$-inch No. 8 screws through the side of the 1x2 and into the floor. Center one hole in the doorway and place the others about 3 inches from each end. Drive in the screws.

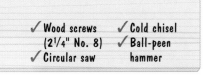

✓ Wood screws
  (2¼" No. 8)
✓ Circular saw
✓ Cold chisel
✓ Ball-peen
  hammer

Prying up wood flooring
Set the cutting depth of a circular saw to the thickness of the flooring minus ¹/₁₆ inch. Make closely spaced straight cuts the length of the floor. Angle a cold chisel into a cut and pry out the strip. Remove all the strips.

DEALING WITH EXISTING FLOORS

How you prepare the room depends on the flooring and subfloor you already have. If your existing surface floor is warped, broken, or damaged, it's usually best to remove it first. Repair the subfloor below if necessary. The chart below describes some common flooring situations. It might help you decide whether laminate flooring is the right way for you to go.

EXISTING FLOOR/SUBFLOOR	WHAT TO DO?
Wood flooring over concrete	Remove wood Clean and smooth concrete Test for moisture Lay down vapor barrier and underlayment or 2-in-1 foam
Flooring other than wood over concrete	Test for moisture Lay down vapor barrier and underlayment or 2-in-1 foam
Any floor over wood	Lay down underlayment
Carpet over anything	Remove carpet—check with manufacturer Clean and smooth subfloor Lay down vapor barrier, if needed, and underlayment

GETTING THE FLOOR CLEAN

Dirt and debris on the floor create high spots that may telegraph through the laminate. You may not notice them at first, but they can eventually cause cracks or wear spots in the floor.
◆ On concrete, scrape off solid materials, such as adhesive remaining from flooring you've removed, with a floor scraper.
◆ On a wood subfloor, scrape gently with the grain; don't dig in.
◆ Once everything is scraped off, clean the floor thoroughly with the cleaning product that your flooring dealer recommends.

1 5

SMOOTHING THE SUBFLOOR

LAST-MINUTE CHECKLIST

Make sure you have:

✓ 2x4 ✓ Belt sander with ✓ Leveling compound
✓ Level   60-grit paper ✓ Putty knife
✓ China marker ✓ Grinder ✓ Trowel

Laminate flooring needs a smooth and level subfloor. Depending on who you talk to, the definition of these terms varies. "Level" usually means a slope of less than $3/16$ inch over a distance of 10 feet. If the subfloor slopes more, get professional help to fix it. "Smooth" usually means that bumps or grooves on the subfloor are less than $1/8$ inch high or deep.

You don't really have to fix minor problems, but the finished floor will look a lot better if you do. It's worth the extra work. Tape plastic over doorways to contain dust. Wear a dust mask, eye protection, and rubber gloves, as necessary. For more major subfloor repairs, hire a professional.

1

Checking for level and flat
Lay a straight 2x4 on edge on the subfloor. Put a level on top of it. Lift the low end of the 2x4 to get a level reading. Measure the gap to see how much the floor is out of level. Pivoting the board, mark the edges of dips or high spots. Check the entire floor.

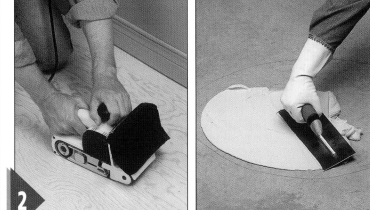

2

Smoothing out the floor
Sand down high spots on a wood subfloor with a belt sander and 60-grit paper _(left)_. Move the tool in the direction of the wood grain. For concrete, use a grinder.

Fill dips in a concrete subfloor _(right)_ with a trowel-on leveling compound. For wood subfloors, choose a product specified for wood. With a putty knife, place some compound in the center of the marked depression. Then smooth out a thin layer of the compound over the area with a trowel and feather the edges. Let the compound dry following the manufacturer's directions.

DEALING WITH MOISTURE

Concrete floors always emit moisture. Each flooring manufacturer allows a specific level; follow their guidelines.

The test at right gives you a rough idea of the moisture emitted from your slab. Generally, if the plastic is dry, moisture is within acceptable levels. Always lay down a vapor barrier—it's required over concrete. Then lay underlayment and install laminate. If, however, the plastic is wet or the floor is damp, moisture levels may be in the unacceptable range. Call a professional for more specific tests and solutions.

Do a thorough outdoor site evaluation to make sure that moisture is not leaking in. Check for crooked gutters and poorly positioned downspouts.

1

Taping down the test plastic

Cut a few 2-foot-square pieces of polyethylene. Duct-tape them to various areas of the subfloor. Wait about 72 hours.

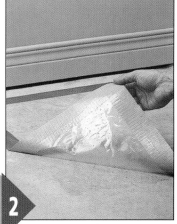

2

Reading the results

Lift up a corner of each test square. Beads of condensation on the underside of any of them, or a dark, moist subfloor indicate a moisture problem. Hire a professional for further testing and solutions.

CRAWL SPACES

Moisture can be a particular problem on floors above crawl spaces. Each laminate floor manufacturer specifies a certain ratio of ventilation per square foot of crawl space. Check that your area meets or exceeds it. If there is a moisture problem and no obvious way to remedy it, have a professional come in to help.

Some manufacturers also require a vapor barrier to be laid on the ground of the crawl space. Even if it's not required, it's a good idea. Install one before moving on.

17

LAST-MINUTE CHECKLIST

Make sure you have:
- ✓ Vapor barrier
- ✓ Underlayment
- ✓ Masking tape
- ✓ Utility knife
- ✓ Tape measure
- ✓ Framing square

Every manufacturer has its own underlayment products and specifications for their installation. Don't try to save money with cheaper substitutes; it may void the warranty on the floor.

Exactly what combination of products you need depends on your situation. At most, you'll need two layers: a vapor barrier and an underlayment. Vapor barriers are always required over concrete subfloors—even if there's a finish floor over top. Vapor barriers also might be required over wood subfloors. Check the manufacturer's directions. Some common underlayments are foam, rubber-backed polyester, and solid panels. See page 6 for product details.

1

Installing a vapor barrier
Unroll the vapor barrier. Butt it against the walls and cut it to length with a utility knife or scissors. Overlap rows by at least 8 inches.

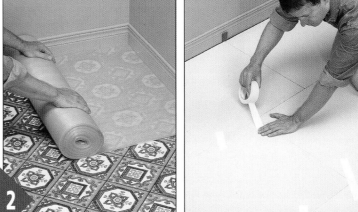

2

Installing underlayment
Butt the end of a foam underlayment against the wall and unroll it *(left)*. Cut it to length with a utility knife or scissors. Most flooring manufacturers recommend installing one row, then covering it with flooring. Butt the rows of underlayment, but don't overlap them. With solid underlayment *(right)*, cover the whole floor. Use a framing square and utility knife to evenly cut the last row to fit. Butt the panels together but leave a ¼-inch gap between panels and the walls or other surfaces. Tape as shown.

LAST-MINUTE CHECKLIST

Make sure you have:

- ✓ Laminate flooring with installation accessories
- ✓ Combination square
- ✓ Pencil
- ✓ C-clamps
- ✓ Saber saw
- ✓ Hammer
- ✓ Compass
- ✓ Damp cloth
- ✓ Straight 1x3
- ✓ Plywood
- ✓ Circular saw
- ✓ Tape measure
- ✓ China marker

Installing laminate flooring is as easy as 1-2-3. Okay, here it's 1 to 16, but it's easier to do than it is to explain. The most important points are:

◆ Use the manufacturer's glue; otherwise, you may void your warranty.

◆ Cut planks carefully. They're durable once they're installed, but they're more delicate than you might think before that. Cut from the back with a circular saw with a carbide-tipped blade or from the front with a saber saw with a laminate blade.

◆ Measure, mark, cut, and install the first three rows precisely. They're the basis of a good installation. The rest of the job will be easier if the first three rows are perfect.

1

Dry-laying the first row
Place the first plank in the corner, grooves against the walls. Put three spacers along the long edge and one at the end. Push the plank and spacers tightly against the wall. Lay as many full planks this way as possible.

Piece to install

Side tongues overlap

2

Marking the cut line
Place the end plank so the end groove butts against a spacer at the wall. With a pencil and combination square, draw a line where it meets the full plank. Continue the line over the edge and across the back.

Wood scrap protects plank

3

Cutting the piece to fit
Clamp the plank to a work surface with the waste side (tongue end) hanging off. Keep the marked line as close to the table edge as possible. Using a circular saw with a carbide-tipped blade, cut along the outside of the marked line.

19

4

Positioning the end piece

Put the end piece in place. Then wedge a pull bar between the plank and the wall and gently tap it with a hammer to ease the plank into place and ensure a tight joint. Slip spacers between the wall and plank. Some manufacturers supply an end clamp, rather than a pull bar, for this task.

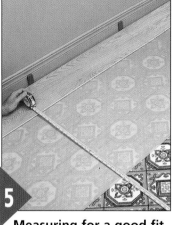

5

Measuring for a good fit

At the middle and both ends of the first row, measure from the edge of the plank to the opposite wall. If measurements are equal, divide by plank width to find the number of full rows you need and how wide the final row will be. If it's less than 2 inches, trim the first row as in Step 6. If the three measurements don't agree, the wall may have a bulge in it; see Step 7, page 21. Also check that the room is square, as described in the box at right.

6

Ripping the planks

If the last row will be too narrow, trim the first row. Measure and mark a cutting line along the back of each first-row plank. Lay one face-down on the work bench with a piece of plywood underneath to reduce chipping. Clamp a 1x3 on top to guide the saw. With a circular saw fitted with a carbide-tipped blade, cut along the outside of the line.

DEALING WITH OUT-OF-SQUARE ROOMS

If your measurements in Step 5 weren't all equal, your room may be out of square. Check by measuring the diagonals of the room. If they're equal, the room is square. Even if they differ by $1/2$ inch or so, the room won't give you any problems. If the difference is greater than $1/2$ inch, use one of the strategies below to deal with it.

◆ Test fit the first row. If the gap along the wall is narrower than the shoe molding, lay the floor as if the room were square. The molding will cover up the problem.

◆ If the gap is too large to be hidden, find a square corner. Begin laying the floor there. Scribe the laminate, as described on page 21, to fit along walls that are out of square.

◆ If the gap is too wide to hide under the molding and none of the corners are square, you'll have to scribe and cut the first row, as described on page 21.

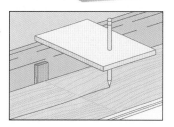
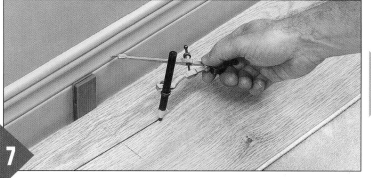

7 Scribing for an uneven wall

Trace the wall or molding outline onto the face of the first row of planks with a compass. If you're trimming around a bulge, set the compass to the widest gap it causes plus about ¼ inch. If you're also trimming because the last row is too narrow, set the compass to the size of the gap plus the amount you need to remove.

Wood scrap protects plank

8 Cutting the plank

Clamp one scribed plank face-up on the table so one end is flush with the edge and the scribed line overhangs by about an inch. Cut along the outside of the scribed line using a saber saw fitted with a laminate blade.

9 Dry-laying the next two rows

Manufacturers don't agree on how much to stagger the joints, so follow the directions for your product. Here, the second row begins with a plank that is one-third its full length. The third row begins with the remaining two-thirds. Dry-lay one row at a time. Fit the joints carefully and cut the end pieces to fit.

10
Gluing the first row

Once you've cut and fit three rows, remove all planks except the first one of the first row. Apply glue to planks as specified by the manufacturer. Here, glue goes in the groove of the second plank. Fit the planks, squeezing the joint tightly. Glue will squeeze out of the joint. Wipe it up with a damp cloth. Glue the rest of the row.

11
Gluing the next two rows

Glue one row at a time, working from left to right. Put glue on the side and edge of each plank, unless they are surfaces that meet the wall. Fit the tongues into the neighboring grooves. Wipe up excess glue.

12
Tapping boards together

Place the tapping block against the plank, fitting its groove over the plank tongue. Tap the block gently to tighten joints. Wipe up the glue that squeezes out with a damp cloth. (If no glue squeezes out, you probably haven't applied enough.) Use the block on both the sides and the ends of the planks.

13
Clamping the planks

Clamp the first three rows together using the clamps recommended by the flooring manufacturer. Space the clamps as recommended—usually two or three clamps per plank. As you place each clamp, wipe up the glue that squeezes out of the joints. Wait an hour for the first three rows to set before removing the clamps and continuing the floor.

14

Installing following rows

Dry-lay, then glue and clamp two or three rows at a time. Start each row with the cutoff from the last plank of the previous row if it's large enough; 12 inches is a good minimum. Stagger joints between rows by at least 8 inches. Tighten the rows with adjustable straps *(above)* or strap extensions. Wipe up excess glue as you go.

Full-width scrap, tongue to wall

15

Scribing final planks

To determine the width of the last row, fit the planks together. Place them face-up, tongues facing the wall, on the second-to-last row. Trace the wall outline on each of the planks, as shown. Cut each plank as in Step 8. Saw to the inside of the line to ensure enough expansion space.

16

Positioning the last plank

Glue the planks of the last row to the planks of the previous row. Then position the final plank and squeeze it to the previous plank and row with a pull bar. Fit spacers at the side and end walls. Wipe up excess glue. Extend straps across the full length of the floor. Let the glue dry before you remove them. Leave the floor to set overnight before walking on it.

UNDER-FLOOR HEATING

Laminate flooring is a good choice over radiant heating systems—hot water pipes or electric heat elements under the floor—which can be too drying for other flooring.

Install the laminate as you would over any concrete floor: Test to make sure the concrete isn't too damp *(page 17)*. Lay a vapor barrier over it—even if it passes the moisture test. Put underlayment on top of the vapor barrier and lay the flooring on top of that.

Most manufacturers suggest that you:
◆ Leave the under-floor heating on for at least two weeks before putting in flooring. Gradually ease the system up to maximum output, then leave it on for about 72 hours. This dries out the concrete.
◆ Turn off the heat and wait at least 48 hours before installing the flooring. During this time, leave the unopened boxes of flooring in the room to acclimate.

WORKING AROUND OBSTACLES

LAST-MINUTE CHECKLIST

Make sure you have:
- ✓ China marker
- ✓ Combination square
- ✓ Laminate flooring and installation accessories
- ✓ Electric drill with Forstner or spade bit

Not every floor is obstacle-free. You'll have to modify planks to fit around obstacles such as pipes and toilets. If you're lucky, you can just drill holes in the plank then slip it over the obstacle, but you're more likely to have to cut it. Whenever you can, remove the object, install flooring under it, and replace it. If you cut the flooring to fit around the obstacle, always leave a 1/4-inch gap for expansion. Fill these expansion gaps with silicone caulk once the flooring glue has dried. Caulk keeps moisture from seeping into the core of the planks. It's especially important in a bathroom or kitchen.

1

Marking the obstacle position on the plank
Install the floor up to the obstacle—in this case, pipes. Make a mark on the front and side of the pipe. Set the end of a plank near the pipes and transfer the center points from the pipes to the plank with a china marker and a combination square *(left)*. Place the plank on top of the last plank you installed, and butt it against spacers, as shown. Draw a line across the width of the plank to mark the pipe centers *(right)*.

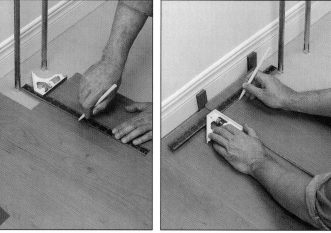

2

Drilling the holes
Clamp the plank face-up on a piece of plywood on a work table. Drill through the center of the crossed lines—using a Forstner bit instead of a spade bit results in a neater hole. The bit must be 1/2 inch wider than the pipes to leave a proper-size expansion gap on all sides.

3

Dividing the plank

On the back of the plank, draw a line through the center of the drill holes. Lay the plank on a piece of plywood on a work bench and clamp it in place. Using a straight 1x3 as a guide, cut along the marked line with a circular saw fitted with a carbide-tipped blade.

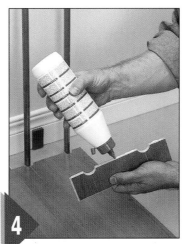

4

Installing the pieces

Place a pair of spacers against the wall behind the pipes. Position and glue the larger of the cut pieces to its neighbors. Fill the groove and cut edge of the smaller piece with glue. Then slip the smaller piece between the spacer and the larger piece. With a pull bar and hammer, tap the pieces together and against the neighboring plank. Wipe away excess glue.

GETTING AROUND LARGER OBSTACLES

Toilets and bathtubs are common large obstacles. (Some manufacturers advise against installing laminate in bathrooms.)

◆ To fit flooring around a toilet, first remove the toilet and plug the drainpipe with a damp cloth to block toxic sewer gases. Scrape the old wax seal off the flange. Dry-lay flooring up to the flange. With a combination square, draw a box to indicate the outer edges of the flange on the plank. Use the same method as for marking pipes *(page 24, Step 1)*. Bend a flexi-curve to the shape of the flange *(top)*. Place the flexi-curve on the plank so its inside edges touch the box. Trace around the outside to lay out a hole with the proper-size expansion gap. Cut along the inside of the marked circle with a saber saw. (The thickness of the flexi-curve creates the expansion gap.) Cut and fit the planks as for other obstacles. Seal around the flange with silicone caulk. When the floor dries, reconnect the toilet.

◆ Treat immovable objects like tubs as you would a normal wall. Start the first row at the tub's edge. Fill the expansion space with silicone sealant *(bottom)*. Smooth it with a wet finger.

FINISHING THE JOB

LAST-MINUTE CHECKLIST

Make sure you have:
- ✓ Tape measure
- ✓ C-clamps
- ✓ Hacksaw
- ✓ Circular saw
- ✓ China marker
- ✓ Electric drill
- ✓ Screwdriver
- ✓ Screws
- ✓ Metal track
- ✓ Molding

Color-coordinated trims—T-moldings, reducer strips, end caps, and so on—make finishing your laminate floor as easy as installing it.

The trim pieces are glued in place or fastened with a metal track. If a track is required, it comes with the screws you need to install it. First, remove the 1x2 backstop *(page 14)*. Then, simply cut the track and molding to length, screw the track in place, and press the molding into the track.

Trim the track with a hacksaw. Cut the molding face-down with a circular saw. Clamp each piece to the work surface so it doesn't move while you cut.

1

Fastening the track
Measure the opening and cut the track to fit with a hacksaw. Butt the ends against the door frame. Place the track so there's a ¼-inch gap between it and the flooring on each side. Mark the track's predrilled screw holes, then drill pilot holes and screw down the track.

2

Positioning the T-molding
Measure and cut the molding to fit snugly against the door frame. Angle the molding slightly so one edge of its underside fits into the track.

3

Pressing the molding down
Angle the length of the molding into the track and snap it into place. Press down on the edges to ensure a tight, even fit.

REINSTALLING SHOE MOLDINGS

Now that the flooring is installed and the transition trim is added, you're almost done. You just need to cover the expansion gap with shoe molding. You can reinstall the old moldings or buy new ones to match your floor.

Installing old moldings is easier because they're already cut to fit. With new moldings, you'll have to cut returns at doorways and archways and miters or coping joints for corners. Coping creates tight joints in out-of-square corners. Practice on scrap pieces.

Drill pilot holes for 1½-inch finishing nails every 12 inches along the molding. Then position it and drive the nails into the baseboard. Protect the floor with cardboard. Countersink the nails and fill the holes.

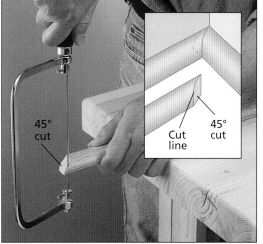

Cutting coping joints
Cut a piece of molding to butt in the corner. Cut the molding you will cope longer than the expanse it will cover—by one molding thickness for each coping joint on it. Cut its end at a 45° angle. With a coping saw, curve the end to fit around the first molding. Follow the line made by the angle cut *(inset)*.

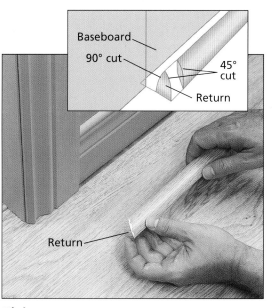

Gluing returns
Cut a 45° angle in the molding that meets the door trim. On a scrap piece of molding, cut a 45° angle in the opposite direction. Then cut a piece off at 90° to create a return. Glue this to the end of the first molding.

DEALING WITH STEPS

LAST-MINUTE CHECKLIST

Make sure you have:
- ✓ Circular saw
- ✓ Tape measure
- ✓ Caulking gun
- ✓ Construction adhesive
- ✓ Nosing
- ✓ Planks
- ✓ Damp rag
- ✓ China marker
- ✓ Clear tape
- ✓ Electric drill
- ✓ Combination bit

When covering steps, forget the idea of a "floating" floor. Here, you glue the planks down and screw the nosing in place, working from the top step down. You may need to trim the last plank at the top of the stairs to make room for the nosing. Trim the underlayment, too. Allow for a 1/4-inch gap between the flooring and the base of the nosing.

Shim up the nosing to match floor height if needed.

Remove any stair flooring that isn't a unsuitable base for laminate *(page 15)*. Saw the overhanging edge of the nosing off flat, fix loose treads, and clean the stairs.

Cut planks and nosing as on page 20; use leftover pieces where you can. Let stairs dry overnight before using them.

WISDOM OF THE AISLES

On some staircases, the underside of the nosing on the upper stairs is visible from the bottom of the staircase. Check your stairs. If you can see the bottom of the nosing, you can still install laminate, but you should make a little adjustment to your stairs first. You need to pad the risers with plywood. Measure and cut a piece of plywood to fit against each riser and screw it in place. Then just install the laminate as described here.

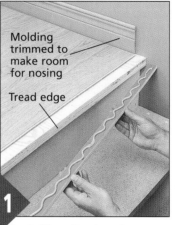

Molding trimmed to make room for nosing

Tread edge

1

Installing the tread edge piece

Measure the exposed edge of the tread and cut a piece of laminate to fit. Apply a bead of construction adhesive to the back of the piece, then press it into place and hold it for a few minutes to let it bond.

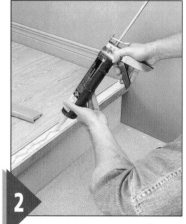

2

Installing the top nosing

Measure and cut the nosing to fit the space. Apply a bead of adhesive to the subfloor, not the nosing. Position the nosing—the tapered end overlaps the flooring—and hold it until the adhesive sets.

✓ Wood screws
(1¹/₄" No. 6)
✓ Screwdriver
✓ Flooring glue
✓ Finishing putty
✓ Putty knife

3

Marking for nosing screws

Starting from the edge of the nosing, measure and mark for one screw every 9 inches. Space holes evenly and center them on the part of the nosing that is glued to the subfloor.

4

Drilling pilot holes

Put a wide strip of clear plastic tape over the nosing. With a combination bit, drill countersink holes for 1¹/₄-inch No. 6 wood screws. Screw down the nosing. Leave the tape in place until after you've hidden the screws with putty *(Step 10)*.

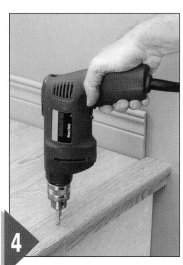

Ripped piece

Full plank

5

Gluing tread pieces together

Measure the stair depth and subtract the nosing depth. If the result is wider than a single plank, rip a second plank to make up the difference. Make the cut on the groove side of the board. Glue the planks together tongue-to-groove.

6

Applying construction adhesive

Lay three beads of adhesive on the tread. Don't put any on the space that will be covered by the nosing.

7 Positioning the tread assembly

Press the glued tread assembly into place on the tread, with the tongue of the full plank facing out. Wipe off any glue that squeezes onto the top of the plank with a damp rag.

8 Covering the riser

Measure the height and width of the riser space. Cut a plank to fit, cutting off the tongue in the process. Apply adhesive to the back of the cut plank. Angle the plank into place, fitting the cut side under the tread overhang of the step above. Press the riser into place.

9 Completing the stairs

Cut a tread edge piece and fit it onto the exposed tread edge, as in Step 1. Press it in place for a few minutes to let the adhesive bond. Repeat Steps 1 to 9 until all the stairs are done.

10 Filling the screw holes

Prepare the putty according to the manufacturer's directions. A scrap of plank makes a smooth mixing surface. With a plastic putty knife, smoothly fill the screw holes in each nosing. Then carefully remove the tape. After about 20 minutes, even out the putty with a cloth dampened with water or acetone. Putty is usually impossible to remove once it's dry, so work carefully and clean up right away.

MAINTAINING THE FLOOR

Laminate flooring is tough and durable, and usually easy to keep clean. Just sweep or vacuum. Damp mop when needed with a bit of ammonia or vinegar in water. Don't flood the floor, as this can cause damage.

Some common products—such as soap, floor polish, scouring powder, and steel wool—are not recommended for laminate flooring. They can damage the wear layer. For tough stains, use the manufacturer's recommended product. Test any products you're unsure of in a less visible area first.

Mats at exterior doorways reduce tracked-in dirt. Felt pads under furniture legs and soft rubber rollers instead of metal or plastic ones help prevent dents and scratches.

Mopping up
Remove stains by spraying them with the recommended cleaner and then mopping up. To keep water from entering the joints, make sure the mop is damp, not wet.

FILLING SMALL DENTS

Because some marks and dents are unavoidable, flooring manufacturers sell fillers for small surface repairs. Most fillers are types of finishing putty, color-coordinated to match the flooring. They resist wear and moisture just like the flooring does.

First clean and dry the area to be repaired. Protect the surrounding area with clear plastic tape. Then fill the dent with putty. Before the putty hardens, wipe away the excess with a cloth dampened with water or acetone, following your manufacturer's instructions. Remove the tape. Dry putty is next to impossible to remove, so clean up well while it's still wet.

More serious damage probably requires replacing planks. Hire a professional laminate-flooring installer.

STUBBORN STAINS

Some stains need more than elbow grease. The chart below offers a few suggestions. Citrus-based cleaners work well for stains that water alone won't fix. Some manufacturers also recommend acetone. Check with your dealer.

STAIN REMOVAL

Stain	Solution
Chocolate	Rub with lukewarm water and a cloth dampened with manufacturer's cleaner.
Tar	Rub with a cloth dampened with citrus-based cleaner or acetone. Wipe with a clean, damp cloth.
Candle wax	Let harden; remove with a plastic scraper. An automobile ice scraper works well.
India ink, paint	Rub with a cloth dampened with citrus-based cleaner or acetone. Wipe with a clean, damp cloth.

Laminate Flooring 1-2-3
Project Director: Benjamin W. Allen
Editor: Jeff Day
Associate Art Director: Tom Wegner
Copy Chief: Catherine Hamrick
Copy Editor: Terri Fredrickson
Contributing Proofreader:
  Margaret Smith
Electronic Production Coordinator:
  Paula Forrest
Editorial Assistants: Karen Schirm,
  Kathleen Stevens
Production Director:
  Douglas M. Johnston
Production Manager: Pam Kvitne
Assistant Prepress Manager:
  Marjorie J. Schenkelberg

Cover photograph:
  Doug Hetherington Photography

Meredith® Books
Editor in Chief: James D. Blume
Design Director: Matt Strelecki
Managing Editor: Gregory H. Kayko
Director, Sales & Marketing, Retail:
  Michael A. Peterson
Director, Sales & Marketing,
  Special Markets: Rita McMullen
Director, Sales & Marketing,
  Home & Garden Center Channel:
  Ray Wolf
Director, Operations:
  George A. Susral

Vice President, General Manager:
  Jamie L. Martin

Meredith Publishing Group
President, Publishing Group:
  Christopher M. Little
Vice President, Consumer Marketing
  & Development: Hal Oringer
Meredith Corporation
Chairman & Chief Executive
  Officer: William T. Kerr
Chairman of the Executive
  Committee: E.T. Meredith III

The Home Depot
Senior Vice President of Marketing:
  Dick Hammill
Project Director: Barbara Koller
Book Development Team
St. Remy Multimedia Inc.
President: Pierre Léveillé
Vice President, Finance:
  Natalie Watanabe
Managing Editor: Carolyn Jackson
Managing Art Director:
  Diane Denoncourt
Production Manager:
  Michelle Turbide
Director, Business Development:
  Christopher Jackson
Editorial
Jennifer Ormston, Stacey Berman,
  Emma Roberts, Brian Parsons,
  Pierre Home-Douglas
Art, Design,
  Illustration, & Studio
Francine Lemieux, Michel Giguère,
  Robert Chartier, Normand
  Boudreault, Jean-Guy Doiron
Production & Systems
Dominique Gagné, Edward Renaud,
  Jean Sirois, Martin Francoeur,
  Sara Grynspan

Special thanks to:
Lorraine Doré, Robert Labelle,
  Karl Marcuse
Consultants
Stewart McLaughlin

Acknowledgments
Active Hardwood Supply, Ltd.,
  Delta, BC
Armstrong World Industries,
  Lancaster, PA
Charles Van Gelder Inc.,
  Albany, NY
Faus Group Inc., Augusta, GA
Formica Flooring, Cincinnati, OH
Global Lumber International Inc.,
  City of Industry, CA
Heatway, Springfield, MO
Mannington Mills Inc., Salem, NJ
Perstorp Flooring Inc., Raleigh, NC
PFG Industries, Vashon, WA
Quickstyle, Montreal, PQ
Quincaillerie Notre Dame de
  St. Henri, Inc., Montreal, PQ
Radiant Panel Association,
  Loveland, CO
Royalty Carpet Mills, Irvine, CA
Tapis Mini-Prix, Longueuil, PQ
World Floor Covering Association,
  Anaheim, CA

Note to the Reader: Due to differing conditions, tools, and individual skills, Meredith Publishing Group and The Home Depot assume no responsibility for any damages, injuries suffered, or losses incurred as a result of following the information published in this book. Before beginning any project, review the plans and instructions carefully, and if any doubts or questions remain, consult local experts or authorities. Because local codes and regulations vary greatly, you should always check with local authorities to ensure that your project complies with all applicable local codes and regulations. Always read and observe all of the safety precautions provided by any tool or equipment manufacturer and follow all accepted safety procedures.

The editors of *Laminate Flooring 1-2-3* are dedicated to providing accurate, helpful, do-it-yourself information. We welcome your comments about improving this book and ideas for other books we might offer.

Contact us by any of these methods:

Leave a voice message at:
  800/678-2093

Write to:
  Meredith Books,
  Wood Flooring 1-2-3
  1716 Locust St.
  Des Moines, IA 50309

Send e-mail to: hi123@dsm.mdp.com